JOHNSON COUNTY PUBLIC L

3 2938 00305

MW01610536

JAN 1 8 1994

J 796.15 G
Goodman, Michael E.
Radio control models.
                              $12.95

WHITE RIVER

Johnson County Public Library
401 South State Street
Franklin, IN 46131

WITHDRAWN

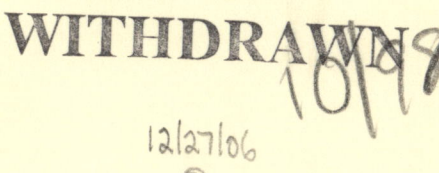

12/27/06

(7)

GAYLORD

RADIO CONTROL MODELS

by

Michael E. Goodman

CRESTWOOD HOUSE

New York

Maxwell Macmillan Canada
Toronto

Maxwell Macmillan International
New York   Oxford   Singapore   Sydney

White River Library
Johnson County Public Library
1664 Library Boulevard
Greenwood, IN 46142

Library of Congress Cataloging-in-Publication Data
Goodman, Michael E.
  Radio Control Models / by Michael E. Goodman. — 1st ed.
    p. cm. — (Hobby guides)
  Includes index.
  Summary: An introduction to radio control models for cars, planes and boats. Information is given on building models from kits and forming R/C clubs.
  ISBN 0-89686-622-X
  1. Automobiles—Models—Radio control. 2. Airplanes—Models—Radio control. 3. Ship models—Radio control. I. Title. II. Series.
TL237.G64   1993
796.1'5—dc20

91-47750

Photo Credits
All photos courtesy of Carstens Publications.

Copyright © 1993 by Crestwood House, Macmillan Publishing Company

All rights reserved. No part of this book may be reproduced or transmitted in any form or by any means, electronic or mechanical, including photocopying, recording, or by any information storage and retrieval system, without permission in writing from the Publisher.

CRESTWOOD HOUSE

Macmillan Publishing Company
866 Third Avenue
New York, NY 10022

Maxwell Macmillan Canada, Inc.
1200 Eglinton Avenue East
Suite 200
Don Mills, Ontario M3C 3N1

Macmillan Publishing Company is part of the Maxwell Communication Group of Companies.

Produced by Flying Fish Studio

Printed in the United States of America

First edition

10 9 8 7 6 5 4 3 2 1

CONTENTS

This radio control model of the offshore racing boat Key Biscayne *is very fast and powerful—and is only 36 inches long.*

THE EXCITING WORLD
OF RADIO CONTROL MODELS

Imagine this: You're behind the wheel of a race car speeding around a track at more than 200 miles per hour. You head into a turn just seconds behind the race leader. You have to manage the turn carefully and not let the car get out of control. That way, you can speed up at the end of the curve to try to overtake your opponent and capture the lead on the straightaway.

Now use your imagination to change your race car into a speedboat. This time, you are skimming across a lake at record speed, kicking up a wave behind you and leaving your opponents in that wake.

Next, mentally transform your speeding boat into a daredevil stunt plane in which you execute exciting loops and dips in the air before bringing it in for a perfect three-point landing.

A race car driver, a powerboat racer, a daredevil pilot—those are all pretty much just dreams if you're a kid, unless . . . you're into radio control models. With a radio control **transmitter** in your hand and a **receiver**, engine and series of **servos** in your model, you can truly be in charge of a car racing around a track, a boat skimming across a lake or pond or a plane or glider flying almost out of sight and back. You can experience the thrills of being in control of a powerful vehicle—and even experience some of the dangers. No, you won't break any bones if you crash a radio control model, but a crash can certainly hurt your pride and your pocketbook.

For thousands—both young and old—radio control (or R/C) modeling is an exciting hobby to begin and to stay with for a lifetime.

Small gliders like this one make good trainers for those who wish to start working with radio control planes.

R/C modelers discover the hobby for many different reasons. Some get the most enjoyment from building realistic scale models of unique planes, boats or cars. They get an extra thrill out of making their models run. Others like speed and power on a small, controllable scale. (Would you believe that an R/C model car recently raced around a banked track at a real speed of more than 200 miles per hour? That was a **scaled speed** of more than 700 miles per hour!) Still others enjoy competing with their friends in contests of speed and skill in the air, on the ground or on the water.

Do any of these reasons appeal to you? Do you have a secret desire to be a stunt plane pilot, a race car driver or captain of your own sailing vessel? If so, R/C modeling may be the perfect hobby for you.

In this book, you'll learn some basics about radio control models. You'll find out how R/C models work, how to build and maintain them, how to race them with your friends, how to join or start a radio control club and get involved in competitions and more. This book will help you learn how to begin the hobby and how to improve on different techniques for building, operating and racing models. It will also provide you with sources for additional information about radio control models.

So, buckle your seat belt, strap on your safety gear and get ready to begin a new adventure that will test your mind and your hands for a long time to come.

A BRIEF HISTORY

Hobbyists have been building model cars and planes since right after the turn of the 20th century. Model boats have an even longer history. But many of these builders had a secret yearning: They wanted to bring their models to life. They longed to be able to operate their models by some form of remote control.

One logical answer was to find a way to use radio waves to control models. This would involve sending radio signals from a hand-held transmitter to a receiver built into the model. The signal could then be converted into mechanical movement by having the receiver send signals to small

The engine of this scale model Corvette is a model. Its hidden electric motor is controlled by radio waves.

electromechanical devices called servos, also installed in the model. Gears inside the servos would be activated and would move **pushrods** connected to the axles or engine of a car; the wings, tail or engine of a plane, or the rudder or engine of a boat. As a result, the modeler could control his or her craft from a distance and adjust its speed, turns or height.

It all sounds pretty simple in principle, but it took a lot of work and imagination from the early R/C pioneers.

During the late 1930s and early 1940s, several different groups of creative and courageous hobbyists finally managed to make their wish to control their model planes, cars and boats come true.

The fathers of radio-controlled flight were two brothers with the appropriate name of Good. Walter and William Good were ideally suited to be R/C pioneers because Walter loved to build models and William was a **ham radio operator**. In 1936, as a college science project, the two brothers combined their interests and constructed a model plane with a wingspan of nearly eight feet (that's huge by today's standards) and powered it across the sky by use of a radio transmitter and receiver.

The Good brothers' efforts didn't immediately inspire others to race into the new hobby. The first R/C planes were heavy, hard to control and very expensive to build and operate. As a result, few people took to the hobby at first. In 1937, for example, when a radio competition was added to the National Model Airplane Championships (NMAC), only six competitors showed up for the event. Just three of them were able to get their craft off the ground, and only one "flyer" could prove that he was in control of his model. He was declared the winner.

During and after World War II, the Goods stayed busy perfecting their models and inventing the basic rules for R/C flying. Their first planes had two controls, a **rudder** for turning and an **elevator** for going up and down. But the Goods believed that rudder control alone was adequate for a beginning R/C flyer. "Complications arising from two controls are apt to be discouraging," they wrote in a magazine article. Even today lots of beginners would agree with that statement. Yet today's R/C flyers nearly all operate at least three controls—rudder, elevator and **throttle** (speed)—and some even add **aileron** (wing tilt) or other specialized controls to their planes, such as retractable landing gear.

R/C cars and boats originated at nearly the same time as R/C planes. The first R/C cars didn't look much like real racing automobiles, however. They were just frames built around radio systems that enabled them to be turned by remote control. Then, a group of hobbyists in southern California, led by a man named Norbert Meyer, hand-made realistic-looking cars that actually raced. Both their speed and turning could be controlled by a human "racer" standing alongside the track. These cars duplicated the thrills of a speedway contest at Indianapolis or Daytona. A new hobby and sport was born!

The first R/C boats were also created in the late 1930s and early 1940s. They were made of timber and powered by steam engines modelers had to build themselves. At first,

R/C boats weren't very popular because of the time and expense required to build them. Inventions such as small internal combustion engines and electric motors, and the creation of lightweight model plastics, have helped to make R/C boating a possibility for many hobbyists.

The Tina *sailboat is large. Besides rudder control, the* Tina *can use a sail winch servo to rotate the sail to the best advantage.*

This large radio control plane is a scale model of a modern Navy jet fighter, the F/A Hornet. When it takes off, the landing gear retracts, just as in the real ones.

R/C planes, cars and boats have come a long way since the 1940s. The guidance systems that controlled these early models transmitted a command which could not be varied in degree. Models could turn only fully left or fully right. Gradual turns were impossible, and movements around a track, for example, could be pretty jerky. In the early 1960s, electronics experts developed radio control systems that allowed for **proportional movement**. Initially, these systems were very expensive, so only a few modelers could afford to equip their craft with them. Then came America's space program and many new experiments with **miniaturization**. These have led to reliable and inexpensive systems that today's modelers can install and operate.

WHAT IS RADIO CONTROL?

You can have lots of fun with a radio control model, but you shouldn't think of it as a toy. It is a sophisticated and fairly expensive machine that requires skill to build and control. It also requires a commitment in time and patience on your part to get the most out of the hobby.

There is a big difference between R/C models and the remote control cars and boats you see advertised in newspapers or magazines for $50 or less. These "toy" models are usually operated by infrared or sonic control and will give only set movements that are not variable. R/C models, on the other hand, use radio waves and a system of equipment that permits proportional movement. Modelers can make their vehicles turn as much or as little as they want and adjust speed in small degrees upward or downward at the same time.

The key to any R/C model is its radio system. After all, control is what the hobby is mostly about. You don't have to be an electronics expert or a ham radio user to install and operate an R/C radio outfit in your model. Proportional radio control equipment is generally available off the shelf at any local hobby shop. Individual components include a transmitter, a receiver, one or more servos, a battery box, wiring harness and an on/off switch. You will have to install everything but the transmitter in your finished model. This takes time and care, but most models come with complete and understandable installation instructions. And hobby store owners or experienced modelers are usually more than willing to help you handle your first radio installation.

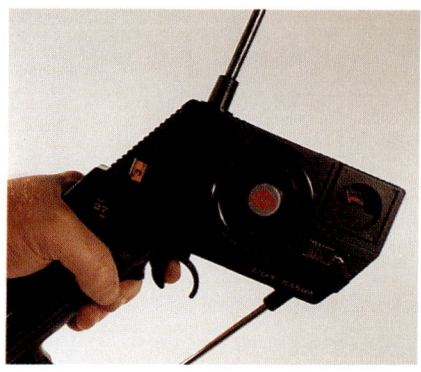

These are two types of hand-held transmitters with which modelers can control their vehicles' delicate movements.

Transmitters come in lots of different sizes and shapes, but they do have some common characteristics. They nearly all have two joystick controls set side by side on the front of the transmitter, with an on/off switch and battery level indicator in between. Alongside or just below the joysticks are small switches called **trim levers**, which help adjust the movement of the servos in your model. There may also be some additional switches or levers if your radio system controls more than three or four functions. An aerial, through which the radio signals will be transmitted, projects from the top of the transmitter. On the back of the transmitter is a small socket in which the radio frequency **crystal** fits. The crystal is usually easy to remove, so that, if necessary, you can change your **frequency** to one not currently being used by someone else on the track, lake or flying field. If you're operating your model on the same frequency as someone else, you might find that the two controls will interfere with each other. Some ways that R/C modelers avoid using the same frequencies will be discussed later in this book.

14

A second vital radio component is the receiver, which is placed inside the model. Most receivers are about the size of a box of matches, but they have several big jobs to perform. The receiver incorporates connections for the various servos in the model; it also contains the battery power supply and an aerial wire. A second crystal of the same frequency as the transmitter also fits into the receiver. This enables the receiver to pick up the transmitter's radio signal and start converting the signal into action.

The major components responsible for this action are the servos. Servos are small electromechanical devices that contain gears and output arms. These arms are then connected to pushrods. When the receiver picks up the transmitter signal, it passes it along to the servos. The servo arms then move in the direction you indicated by maneuvering one of the transmitter joysticks until the servo balances the radio signal. The pushrods, connected to the servo arms by wire linkages, adjust speed or direction of the R/C model plane, car or boat.

This is a typical hookup in an R/C vehicle. It shows a large and medium-size servo attached to the receiver. At right is the battery and the on/off switch. All this goes inside the vehicle.

You're also going to need batteries to start your engine and to power the transmitter and receiver. Nearly all modelers use rechargeable nickel-cadmium batteries, or **ni-cads**, which hold a charge for 20 to 25 minutes and take about 15 minutes to recharge. These batteries can last through more than a year of normal use.

Radio control systems can be simple or complex, depending on how much you want the system to control—and how much you want to spend. If you're planning to run a car or boat, you will need a system that generates two or three signals, called **channels**. One signal will adjust speed, another will control turning and a third can be used for some other function, such as shifting gears or trimming sails. R/C plane modelers need a system with at least three channels—to control rudder, elevator and throttle—but it is probably best to purchase a four-channel radio outfit so that your system can grow to let you also control the plane's ailerons as well.

This is a typical setup found inside a model plane's fuselage.

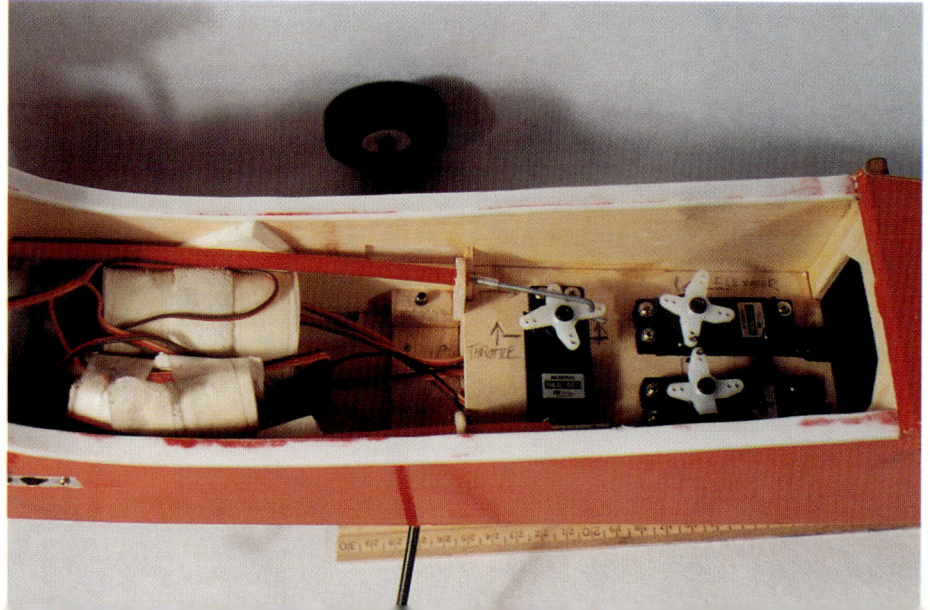

This high-performance racing boat features what is called an outdrive—a combination rudder/propeller that steers and turns the boat.

The best advice is to start simple. In fact, the motto of most R/C modelers is "KISS," which stands for "Keep It Simple, Stupid." Start with only those functions you need and can control comfortably. Then build your system gradually as your skill and interest increases. R/C modeling is a hobby that can last a lifetime. Don't rush ahead too fast, or you might become discouraged and quit before the real fun begins.

CHOOSING YOUR FIRST R/C MODEL

Once you have decided to get into R/C modeling, you have a lot of questions to ask yourself. For example:

Should I choose a car, boat or plane for my first model?

That choice obviously depends upon which interests you the most. But many hobby store owners and veteran R/C modelers suggest that young modelers begin with cars or boats. These require you to think only in two dimensions—turning and speed. And even those two dimensions can present quite a challenge. It's a mind-bending task to maneuver a car's wheels through a turn and also think about adjusting the engine speed so that the turn can be made under full control without flipping your car over. Planes add a third dimension—height—and an added risk, the possibility of crashing and destroying an expensive model. As you might expect, R/C model planes are also generally more expensive to purchase and maintain and more complex to build than either R/C cars or boats. They also require more lessons and patience to master.

You can operate an R/C car or boat anywhere there is an open, flat surface or small body of water. R/C flying requires a large, open area, free of trees, buildings, electric or phone lines and people. For space and safety reasons, you'll probably also have to join a club to be an R/C flyer. Clubs can be fun for sharing your interest in R/C cars or boats, but they aren't a requirement. These issues will be discussed further in later chapters of this book.

So there are no easy answers to your first question. You have to examine your own interests, personality, surroundings and pocketbook to help you decide what type of model to begin with. Each modeler probably has a different explanation for his or her choice. Two teenage R/C car hobbyists from Matawan, New Jersey, Ed Pisani and Paul Odell, said they chose cars for three main reasons: They could run their models in their backyards; they could build models quickly and get the thrill of control and speed for a reasonable cost; and they couldn't stand the idea of taking out an expensive plane for the first time and crashing it.

Two R/C cars racing. The long, flexible white shaft is the radio's antenna.

Another teenage modeler, Jim Bowen, from Tempe, Arizona, said he started with cars but quickly moved to planes. "The cars just didn't hold my interest," he said. "Cars do the same thing over and over again. A plane presents a new adventure each time up."

A fourth modeler, Fran Crowne, of Clayton, Missouri, said she chose R/C boating because of the wide variety of model types available—from racing hydroplanes to battleships to paddle wheel riverboats to sailboats. She also found running her models more relaxing than operating either a car or a plane. "My boat floats, so I don't have to worry about keeping up a certain speed to run it. And I have lots of time to correct for errors operating the radio controls. I'm not worried that my model is going to crash any second."

How much patience do I have for intricate detail in building a model and how quickly do I want to feel "in control" of my model?

If you're the impatient type, start with a car or boat. A simple car kit can take as little as three to four hours to assemble, including the radio installation, and may require reading only a page or two of directions. On the other hand, a basic R/C **trainer plane** might come with eight to ten pages of directions and can take 20 or more hours to build.

What kinds of models are best for a beginner?

Whether you choose a car, boat or plane, begin with something that isn't too complex or expensive to build or control. For starting car modelers, that usually means a basic van that doesn't move too quickly (perhaps 20 to 25 miles per hour), is solidly balanced and contains plenty of space for the radio control equipment. Some of the same characteristics should be true of your first R/C boat. Most new R/C boaters don't start with a sailboat unless they already understand basic principles of sailing full-sized boats. You'll probably only need a two-channel radio system to begin with in either your first car or boat.

For beginning modelers another decision is whether to choose a model with an electric engine or a gas one. One hobby store owner's advice is: "Go with electric. It's quieter and easier to get running right. Gas engines have to be installed absolutely correctly or they are hard to start."

Why do many modelers go with gas? Gas engines may be noisier, heavier and more expensive than electric ones, but they are also faster. If you're into speed and have the patience to get it running right, you might choose a gas-powered model.

Beginning R/C pilots usually start with a trainer plane and a three- or four-function radio system. The wings on these trainers fit atop the body of the plane. That way they are farther from the ground and less likely to be damaged in case of a crash. The wings are also flat on the bottom. This helps the plane to be more stable. In case the craft goes out of control, the flat bottoms of the wings will enable air current to push the plane out of a crash pattern and give the pilot more time to even out the flight of the model. Trainers usually don't contain ailerons, so a three-function radio system is all you will need to control speed, turns and height.

How can I find out what models are available?

The first place to go is your local hobby store. Talk with the owner there or some of his or her customers. These people are usually experienced modelers, and they'll probably welcome your interest and love telling you about their modeling "adventures." You can learn a lot from listening to the stories they tell about their successes and failures as beginners. Also, look at the types of kits available to see how different and complex they are. Most of all, don't be afraid to ask questions.

You can also pick out one or more hobby magazines, such as the ones listed in the back of this book. These contain interesting articles about modeling or radio control, as well as lots of ads for different models and equipment that are currently on the market.

You might also find out from the hobby store owner where R/C car, boat or plane clubs in your area meet. Go and visit one or more of these and watch different modelers and their vehicles at work. Find out what brands or types of vehicles or radio systems the members of the clubs use and recommend or suggest you avoid. You might also find a good "teacher" at one of the clubs who will help you build and learn to operate your first model.

How much will I have to spend for my first R/C model?

This is another difficult question to answer because there is such a variety of model cars, boats or planes available and prices change from year to year. At the time of this writing, a basic R/C van or car could be bought for around $100, with a simple radio system running $60 to $120 more. Basic boat prices are similar, but a fancier boat can run well over $200 for the model alone. A basic trainer plane might run around $100 to $150, with a radio system nearly matching that price. If you want something fancier, "the sky's the limit."

There are also different accessories, tools, electric-engine starters, batteries and battery chargers, spare parts, fuel, wire, paints, coatings and adhesives that present additional costs to consider. You'll probably also need to invest in a box to store and transport all of your equipment conveniently. You should check magazine ads or hobby store shelves for the latest pricing information on all equipment and accessories.

One piece of advice: If you plan to stay with R/C modeling and can afford it, don't start with the least expensive model available. You'll probably outgrow this type of model quickly and then want to move up to something a little better constructed and more interesting. So you may end up buying two models instead of one. Ask your hobby store owner or other experienced modelers for ideas of the best models both to start with and stay with.

Simpler R/C cars, like this Madcat, are two-wheel drive. They are used as entry-level cars because they don't have extra features and they're fairly inexpensive.

KITS VS. ALMOST-READY-TO-RUN

Part of the issue of choosing your first R/C model is deciding how much or how little building you want to do. For some R/C modelers, planning and putting a model together from scratch is the most enjoyable part. These modelers either develop their own design or follow plans found in a magazine or book. They construct their cars, boats or planes from a variety of materials such as **balsa** wood, plywood, foam or plastic card. Scratch modeling takes a lot of time, effort and care, but it's a lot of fun too. And at the end, you have a product of your own hands to show off.

Other R/C hobbyists are more interested in running and racing their models than in building them. They're willing to spend more money to buy fully assembled models and need only to add radio control equipment to get their vehicles up and running.

However, most beginning modelers start somewhere in between these two extremes. They purchase kits that are completely or partially unassembled. The quality of most kits is excellent, and everything is supplied except the radio control equipment. Some kits also include an electric motor. Since most parts are prefabricated, assembling a model from a kit is a fairly direct process—*if you follow the directions carefully!* There is nothing more frustrating than putting everything together on a model and then finding a few parts left over on your worktable.

Most beginners choose kits in which the parts are already partially assembled. These are called **almost-ready-to-run** models. In the case of planes, this type of model is called an **ARF**. That's short for almost-ready-to-fly. But don't be hoodwinked into thinking that almost-ready-to-run or almost-ready-to-fly means you'll be out on a track, lake or flying field in a few minutes. Even partially assembled kits take time and care. Rushing may lead to more work rebuilding things than if you did it right the first time around. Figure at least 3 to 5 hours to construct a partially assembled car, twice as long for a basic boat, and from 10 to 20 or more hours for a basic ARF.

An almost-ready-to-run car kit requires just a few common tools to assemble. You'll need a screwdriver, wrench and pliers. The best investment you can make is an electric screwdriver. Not only will it save you time and effort, it will also help you avoid damaging your model by using too much pressure by manually tightening screws. Here's another helpful hint: Put soap on your screws before you tighten them. They'll go in faster and easier.

If you are careful and follow the directions precisely, you might find that painting your almost-ready-to-run model car will take more time than putting it together.

Almost-ready-to-run boats take more effort to assemble than cars, but less than planes. There are propellers, flywheels, shafts and couplings that must be applied carefully. You may also need to do some carving of pieces of the wood or hard plastic hulls so that they fit together properly.

Another problem you might incur is that the hardware included in many boat kits—steel wire for shafts, brass tubing, etc.—is often too simple and flimsy for your needs. You might want to replace them with better quality materials before you get started.

ARF model planes are usually already painted, and the special heat-sensitive Mylar coating needed to protect the wings during flight and landings has been applied by the manufacturer. (With a completely unassembled model, you'll need to apply the coating yourself. This requires relatively simple but careful work using a heat iron for stretching the coating and a heat gun for sealing it. You can also use a fabric coating, such as silk or tissue, and apply it to the wings.)

Assembling and aligning the various parts of an ARF must be done with great care. Parts must fit together snugly to avoid vibration or separation while in flight. Improperly aligned wings can also cause your plane to be hard to get aloft and to control in the air. One good practice is to test-assemble and align parts of your model before gluing them together. Push the parts tightly together until little or no gap exists between them. You might need to do a little sanding so that adjoining parts touch together everywhere. Rubber-band the wing and **fuselage** together and measure from the fuselage to each wing tip to make sure the distance is identical. Follow the same process to check tail alignment before you apply the hinges.

To construct an ARF, you'll need different types of fast- and slow-drying adhesives. Fast-drying cements are good for parts that don't have to be aligned exactly. Slow-drying cement gives you time to manipulate parts that must fit together perfectly. Some parts, such as the rudder, also need to be hinged to the body of the plane. This is another gluing step.

Before starting, gather all the materials you'll need. You should have glue, solvent, solder, sandpaper, a ruler, drill, awl and possibly a vise on hand. You also need something to apply the glue. Long cotton swabs or strips of balsa wood make great glue applicators for hard to reach places. Then do your test assembly and measurements before you mix the epoxy. Once the glue is mixed, you'll only have between 15 to 45 minutes to use it before it dries. That's why planning ahead can save you lots of time and headaches in the long run.

For those who don't want fast or fancy R/C boats, there are R/C tugboats like this Bugsier Harbor tug. It can tow barges just like the real thing.

No matter whether you build your model from scratch, purchase an unassembled kit or choose an almost-ready-to-run model, you'll still have to install the radio control equipment and engine yourself. That's a difficult task because you have to mount a lot of moving parts—receiver, servos, pushrods, etc.—into a limited space. But radio installation is not an impossible chore. Plans included with better kits show side and top views of proper installation and generally take you step by step through the process. There are also many books on radio and engine installation available at your hobby store. Even so, you're probably going to need some help installing parts into your first R/C model. Remember the advice you were given in the last chapter: Don't be afraid to ask questions. You might also check with the hobby store owner or experienced modelers to find out what special tools you can purchase to make your task easier. One such tool that can save endless time is a pair of z-bender pliers. These can ease the task of bending hard piano wire into the proper shape to attach servos or engines to your model. These special tools aren't absolutely necessary, but they can certainly help.

All of the information in this chapter should make it clear that there is an R/C model kit available to meet almost every modeler's interests and desires. And for the more creative or courageous hobbyists there is a large variety of raw materials for building from scratch. What's right for you? Deciding is part of the fun and challenge of the hobby!

RACING ON A BUDGET

R/C modeling requires two kinds of investments on your part—time and money. There is no getting around either one. You can, of course, save time by buying a completely assembled model or by hiring someone to help you build your new car, boat or plane. But that means you'll be spending more money and missing out on a lot of the fun of the hobby. Almost-ready-to-run kits also save you time and effort. For beginners this savings is usually worth it. Many beginners just don't have the patience or skill to put together a completely unassembled kit the first time out.

Saving money is another issue. Contrary to what you might think, an unassembled model kit is not necessarily less expensive to put together than an almost-ready-to-run one. You will need more tools and materials to complete your assembly and may find that you will spend almost the same amount of money either way.

There are other ways to cut expenses, however. The first way involves the type of model you choose. If you look through an R/C model magazine, you will see lots of ads and catalog listings that include prices for kits, radio control equipment and engines. Remember that the most expensive models are not necessarily the best either to assemble or to run. Even if a certain model is "hot" today, its popularity may soon fade. You want to choose a model that you can stay with for a while and that will grow with you as your skills improve. Also, decide where you're planning to operate your model. A car that you're planning to run in your backyard or on your driveway will probably be a very different one from the car you plan to race on a track. The same is true of boats or planes. Discuss your needs and the differences between the various models with your hobby store owner. The owner can help you pick one to match your interests, skills and budget.

Here are a few more hints for modelers with a tight budget:

(1) An electric car or boat may run slower than a gas-powered one, but it is also less expensive both to purchase and operate and easier to get running properly. It will also require a smaller investment in mufflers to reduce the noise gas engines create.

(2) Radio control equipment also varies greatly in price, largely dependent on how many channels you need to control. For most basic cars and boats, a two-channel system is all that's required. Buying a larger outfit than you need can unnecessarily push up your costs. On the other hand, a four-channel system is usually less expensive than a three-channel one. So if you're buying control equipment for a model plane, even if all that's required are three channels, you might want to invest in a larger system. Besides, the extra channel may come in handy if you decide to add ailerons to your current or next model.

(3) You might want to share the cost of some of the necessary "extras" with a friend. These might include battery chargers, electric starters, extra radio crystals for different frequencies or tools needed for constructing models.

This four-wheel-drive monster truck has real coiled spring shock absorbers.

Radio control modeling can be a fairly expensive hobby. But don't let a tight budget keep you from enjoying it. And don't opt for the cheapest model or equipment if you're planning to stay with the hobby. As discussed earlier, pick a model that will grow with you. Otherwise, you may get bored and give up the hobby or want to invest quickly in a more sophisticated model.

LEARNING TO DRIVE, SAIL OR FLY

R/C models are fun to build, but they're even more fun to operate. An R/C model has an advantage over a toy car, boat or plane because you have the ability to control its speed, direction, turns or loops. But don't expect to grab a transmitter and be in full control right away. You will need to apply some of the same patience you showed in building your model to learning how to operate it properly. You will probably also need to ask for help from a friend or an experienced modeler. Lessons are particularly necessary for a beginning R/C pilot. You have too much time and money invested in your model plane to crash it on the first time out because you are too shy or stubborn to ask for help.

As you might expect, R/C cars are usually the easiest to learn to control. It takes most new drivers only a few runs to get used to being "behind the wheel" of an R/C car. One of the joysticks on the transmitter controls the car's speed and gears. You move it up and down for forward, reverse and throttle. The other joystick steers the car. You push this stick in the direction you want the car's wheels to move. That is pretty simple as long as the car is moving away from you. But if the car is coming toward you, turns must be made oppositely. For example, pushing the stick to the right will move the car toward your left. Most new R/C drivers find this a little tricky at first.

Speed presents another problem for an inexperienced R/C driver. Remember that your car is moving in "scale" speed. In other words, a car one-eighth the size of a life-size model with a comparable engine will act as if it's moving at eight times its real speed. As your car moves into a steep turn, you'll have to slow it down, or it might flip over or race out of control. Even experienced R/C drivers have trouble controlling speed. That is why, during a race, marshals are usually posted every ten feet or so to flip cars back over.

R/C motorboats present some of the same problems as cars for an inexperienced captain, who must learn to control direction and speed. R/C sailboats are a different story, because in addition to other concerns, wind currents and sail positioning must be considered. Make sure that your first few sailing expeditions are made with an experienced R/C captain at your side.

R/C planes and gliders are a real challenge to learn to operate properly. The most obvious reason why planes are more difficult to run is that a pilot has to concentrate on three separate dimensions—direction, turns and height—as well as model and wind speed. In other words, while you are controlling the forward movement of your plane, you also have to concentrate on keeping the nose of the plane up, compensating for wind direction and making turns. In addition, because even a large R/C plane may appear to be only a speck when it is far from the pilot, it isn't always easy to tell whether a model is moving away from you or toward you. Therefore, you are not always certain which way to push your rudder joystick to make the proper change of direction. You also don't usually have as much time to correct for mistakes with a plane as with a car or boat. If your mind goes blank for a few seconds, your plane may sail out of sight or head dangerously toward the ground.

Luckily, you'll probably have little trouble finding someone at a hobby store or on a flying field who is willing to be your instructor. Some trainers use a special set of controls called a **buddy box** to help beginners learn to fly. A buddy box contains two transmitters that can be used to control the same plane. As long as the beginner is doing okay, he or she controls the craft. But if a threatening situation occurs, the trainer can take over for a few seconds until the beginner is ready to take charge again. Other trainers don't believe in buddy boxes. They would rather talk a beginner through problems, so that the new pilot learns directly from his or her mistakes.

A beginning R/C pilot needs a lot of patience and perseverance. You're not going to learn it all the first time out. And changes in wind direction, cloud cover and atmospheric conditions may make every new day of flying a new challenge. This adds to the frustration of learning to fly but also increases the fun of meeting the challenge.

FINDING A PLACE
TO RUN YOUR MODEL

There are certain rules both of courtesy and safety that must be followed in determining where you should run your model car, boat or plane. Some of these were discussed earlier, in the chapter on choosing your first model.

Safety is the first requirement. You don't want to do anything that will harm your model or another person's model or that will endanger spectators. So make sure the place you choose is free of obstacles. For a model car that means a flat surface or one free of potholes, rocks or limbs.

Car races are usually organized by clubs, because they require a large, open, paved or carpeted area with a marked-off course. The course has to be wide enough to accommodate several cars racing at the same time and must have room for the drivers to stand near each other and operate their vehicles. Wooden barriers are often used on the outside areas of the track to prevent runaway cars from crashing into walls or spectators. High-speed races are sometimes held on banked tracks to keep cars from flipping over.

How can you find out about clubs or tracks near where you live? Once again, your best source of information is a local hobby store. Many hobby stores help their customers organize clubs and set up race courses. In fact, the parking lot behind a hobby store often doubles as a track on racing days.

If you're sailing an R/C boat, you'll want a body of water that's clear of obstacles and not being used by swimmers or flocks of birds. It's probably best to have a place with little or no current to carry the boat out of your control and one that is enclosed and doesn't lead into a larger stream or lake.

R/C pilots require lots of clear space away from trees, wires and buildings to fly their planes.

Noise is another important factor for both car and boat operators. R/C vehicles make a lot of noise, even when proper mufflers are installed. You'll probably need to find a place that is not in the middle of a residential area to help avoid complaints from annoyed neighbors.

R/C model airplanes require lots of clear air and land space as free of trees, wires and buildings as possible. Finding a large, safe flying field is one reason nearly all R/C pilots belong to flying clubs. A flying club usually reserves space during certain times or days and maintains the area. The space is often as far from main roads or residential neighborhoods as possible. Believe it or not, a truck using a CB radio while driving along a nearby highway might cause someone to lose control of an R/C plane.

A second reason for belonging to a flying club is control; the club regulates radio frequencies being utilized at the same time to avoid interference and possible accidents. Flying clubs maintain a board on which frequencies currently in use are posted, or pilots attach special colored flags to their antennas to indicate their frequencies. If your radio crystals match the frequency of another pilot, you'll have to wait your turn to fly.

A third reason behind flying clubs involves insurance. Planes occasionally do go out of control and do crash into buildings or people, causing damages or injuries. You're going to need **liability insurance** just in case of such an accident. Flying-club members all belong to the Academy of Model Aeronautics (AMA), and a large liability insurance policy is included in the AMA membership fee. The AMA's address is listed in the back of this book.

JOINING OR FORMING
A RADIO CONTROL CLUB

You have just read about some of the reasons that R/C pilots belong to flying clubs. But the main reason for joining a club devoted to R/C planes, boats or cars is the opportunity to be with other people who share your interest. In a club, you can learn what new equipment to buy or avoid based on the experience of other modelers who have already put it to use. You can discover new techniques for building or operating your models. You can find friends to help share the cost of some of the "extras" that can make building or running models easier.

Are there clubs in your area? Your hobby store is your first stop for information. You can also check some of the magazines or national organizations such as AMA, ROAR (Radio Operated Auto Racing Association) or NAMBA (North American Model Boating Association) listed at the end of this book.

If there is no club in your area, consider forming one. Put an ad in your school newspaper or post an announcement at your library or hobby store. The hobby store owner will probably be happy to help you advertise and form your club and may even supply the space for meetings. But don't count on him or her to do all of the work setting up meetings or conducting competitions. After all, it's your club, and the hobby store owner also has a business to run.

How can you get a club off the ground? Call an organizational meeting. Show a film on racing sports cars or boats or flying planes and display a few R/C models to attract attendees' interest. You might even invite someone in your area who races full-sized cars, sails a full-sized yacht or flies a small plane to be a guest speaker at your first club meeting. Then quickly ask attendees to vote on a name for the club and hold an election of officers. This will make the club seem more official and may keep people coming back for additional meetings.

Once the club is functioning, its members can help organize races and other competitions, modeling exhibitions or training sessions for new modelers.

Joining or forming an R/C club can help you meet other modelers and share ideas, techniques and fun.

39

Competitions offer exciting ways to test your abilities as a pilot.

RADIO CONTROL COMPETITIONS

Most R/C clubs and national organizations sponsor various competitions so that modelers can test their skills at building and running models with other hobbyists. For example, once a year, ROAR organizes the National Road Racing Championship, where a grand national champion is chosen, as well as national champions in such categories as drag racing, Indy-style racing (around an oval in a left-hand turn pattern) or road racing (around a closed course that includes left and right turns of varying sharpness). NAMBA and AMA also hold national competitions. Boating champions are crowned in speed and sailing categories. Model flyers vie in glider (length of time in the air and accuracy of landing) and model plane competitions (timed races around pylons set in a triangular pattern or test of ability to execute aerobatic maneuvers and patterns). All three organizations also sponsor competitions in scale modeling—how well a model duplicates a full-sized vehicle on a set smaller scale. These competitions are designed for modelers who are more interested in building than in racing.

In addition, local R/C clubs often hold races on weekend afternoons on a large parking lot, pond or flying field. The clubs usually notify the national organization that they plan to hold a meet. The national organization then sanctions the meet (makes it official) and helps handle publicity and insurance coverage. The competitions must follow standardized rules established by the national organization. That way, a competition in South Dakota has the same standards as one held in Maine or Alabama.

But you don't have to be a contestant to enjoy one of these competitions. Watching and listening may be a good first step to take before you even decide to get into R/C modeling. You can see different types of models and equipment in use and talk with experienced or beginning modelers.

But here's a warning: The enthusiasm of the modelers you see during a competition is probably going to be contagious. You may find yourself ready to visit a hobby store on the Monday following a Sunday competition that you have observed. If you do, then welcome to a hobby that is fun to begin and fun to stay with for a lifetime.

FOR MORE INFORMATION

This book is only a starting place. There are lots more sources of information you should explore as you learn about R/C models and try your hand at the hobby. Here are some good places to turn to for more information:

ASSOCIATIONS & ORGANIZATIONS

The Academy of Model Aeronautics (AMA)
1810 Samuel Morse Drive
Reston, VA 22090

This is the major organization for R/C model plane hobbyists, and even many R/C boaters or auto racers belong. Members receive a listing of clubs in their area and regular issues of Model Aviation *magazine. The AMA also publishes standards for the hobby and provides a $1 million liability insurance policy as part of its membership fee. It holds an annual meeting in association with the National Model Airplane Championships, which have been held since 1936.*

There are two organizations for those interested in R/C gliders:

The National Soaring Society
3755 Berkley Lane
Lumberton, NC 28358

League of Silent Flight
P.O. Box 647
Mundelein, IL 60060

Radio Operated Auto Racing Association (ROAR)
3703 Dover Drive
Fort Wayne, IN 46805

This is the major organization for R/C car modelers. It serves as a central governing body for the hobby, sanctions competitions and publishes sets of standards. In addition, members receive a monthly newsletter with lots of shopping and construction hints.

North American Model Boating Association (NAMBA)
1815 Halley Street
San Diego, CA 92154

This is the major organization of R/C model boating. It sanctions races, holds competitions, presents awards and maintains a hall of fame for outstanding modelers and racers and their designs. It also publishes a monthly newsletter, Propwash.

For model sailboaters, the following organization may be of interest:

American Model Yachting Association (AMYA)
1884 Campus Court
Rochester Hills, MI 48309

PERIODICALS

Below is a list of magazines that will appeal to those interested in R/C models. Most issues contain interesting and useful ideas for building and racing models as well as scores of ads and mail-order pricing catalogs.

R C Modeler
R C Modeler Corporation
144 West Sierra Madre Boulevard
Sierra Madre, CA 91024

Flying Models (also incorporates R/C Boating)
Carstens Publications, Inc.
P.O. Box 700
Newton, NJ 07860

Air Age
Air Age, Inc.
251 Danbury Road
Wilton, CT 06897

Radio Control Car Action
Air Age, Inc.
251 Danbury Road
Wilton, CT 06897

Radio Control Model Cars
898 West 16th Street
Newport Beach, CA 92663

BOOKS

A variety of general books on R/C car, boat or plane modeling and those focusing on specific subjects such as construction, radio control installation, electronics, engines and racing are available from the following publishers. You can find many of the books in your public library or in hobby shops, or you could contact the publishers directly at the addresses listed below:

Air Age, Inc.
251 Danbury Road
Wilton, CT 06897

Carstens Publications, Inc.
P.O. Box 700
Newton, NJ 07860

Harry B. Higley & Sons, Inc.
P.O. Box 532
Glenwood, IL 60425

Kalmbach Publishing Co.
21027 Crossroads Circle
P.O. Box 1612
Waukesha, WI 53187

TAB Books
Blue Ridge Summit, PA 17214

GLOSSARY

aileron—The movable portion of the wing of a plane which controls tilting movement of the aircraft.

almost-ready-to-run—Partially constructed kits for model cars or boats.

ARF—An abbreviation for "almost-ready-to-fly," which is used to describe a partially constructed model airplane kit.

balsa—A soft, lightweight wood often used for building model planes, gliders or boats.

buddy box—Dual-control transmitter sometimes used to help train a beginning R/C pilot by enabling both the instructor and student to control the model plane.

channel—A one-way flow of radio waves sent from a transmitter to control one function of a model vehicle, such as turning, speed or elevation.

crystal—A small, rectangular piece of quartz that vibrates strongly at a certain frequency and that is used to regulate the frequency of radio waves sent from a transmitter to a receiver.

elevator—The control that enables an R/C pilot to adjust the height of the plane.

frequency—Radio wave signals of a certain pattern that are sent from the transmitter to the receiver. Each vehicle operates on a specific frequency that is determined by the crystals used.

fuselage—The body of a model airplane.

ham radio operator—An amateur radio operator who uses certain frequency bands for communication.

liability insurance—Policy that protects an R/C modeler against responsibility for paying for damages or injuries caused by a model's crashing into something or someone.

miniaturization—The process of producing an exact working copy of something in a reduced scale.

ni-cads—Rechargeable nickel-cadmium batteries used to power transmitters, receivers and electrical engines of models.

proportional movement—Movement of a model that directly matches the amount of movement on a control stick of a transmitter.

pushrods—Rods connecting the servos to the parts of an R/C model that they will move.

receiver—Part of an R/C unit that receives the radio signal sent from a transmitter, decodes the signal and then sends it on to the appropriate servos.

rudder—The control that enables the R/C pilot to turn the plane.

scaled speed—The speed of a model's movement measured in the same ratio as the model's size to that of the full-sized vehicle it copies.

servos—Small electromechanical devices outfitted with rods. The rods move various parts of an R/C model in response to signals sent from the transmitter and conveyed by the servos mounted in the receiver.

throttle—The portion of an R/C model's engine that adjusts how quickly the engine operates and thus controls the model's speed.

trainer plane—A sturdy, easy-to-handle R/C model plane used by beginning hobbyists.

transmitter—The part of an R/C unit that sends radio signals to a receiver mounted in the model.

trim levers—Small switches located on a transmitter that adjust and balance an R/C model's servos.

INDEX